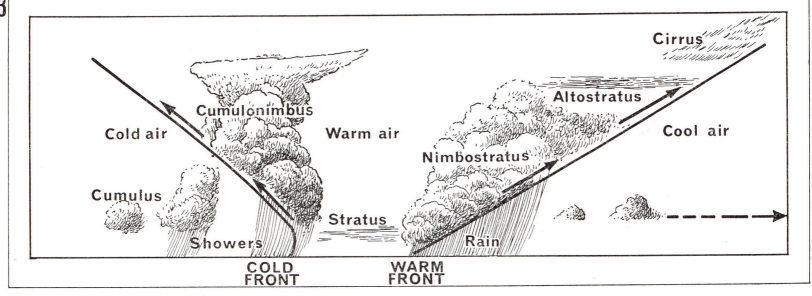

Section through a depression. End-paper No. 3 shows a section through the warm and cold fronts of a depression which is moving in the direction of the broken arrow. At the warm front the warm air is rising over the cool air in the direction of the arrows. The slope of the frontal surface is from 1 in 150 to 1 in 300. High cirrus and probably cirrostratus cloud form as much as 500 miles or more ahead of the warm front and are the first signs to an observer on the ground of the approaching front. Nearer to the front the cloud becomes lower and thicker, and there is a belt of precipitation (rain or snow), often about 200 miles in width, falling from nimbostratus cloud just ahead of the front. In the warm sector the temperature rises and some drizzle may fall from the low stratus cloud. At the cold front the cold air pushes like a wedge under the warm air, which rises in the direction of the arrows up a slope of about 1 in 50. Here there is cumulonimbus or heavy cumulus cloud, giving a shower, with perhaps thunder, and probably followed by further showers. In the air behind the cold front the temperature is lower than in the warm sector. If you study the plan of a typical depression (end-paper No. 2), you will be able to work out the changes in wind direction and atmospheric pressure that take place at the warm front and the cold front. The wind veers at both fronts; the pressure falls ahead of the warm front, becomes steadier at the front, then rises suddenly at the cold front.

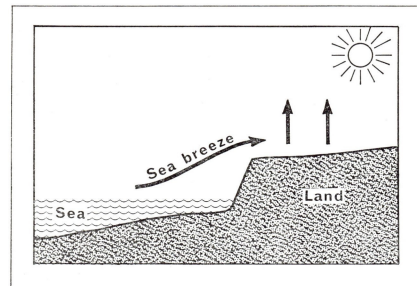
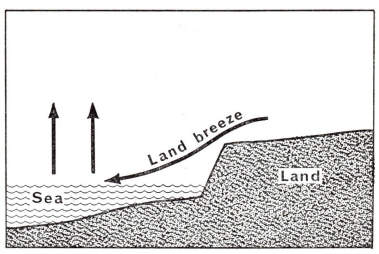

Sea breeze and land breeze. End-paper No. 4 shows the formation of two local winds which are important in many coastal districts of the world: the sea breeze, which blows during the day, and the land breeze, which blows at night.

Sea breeze. During the day the sun warms the land more than the nearby sea. The air over the land is heated and rises, so that the atmospheric pressure becomes lower than it is over the sea. Air above the surface of the sea therefore flows towards the land, and this moving air is called a *sea breeze*. In temperate regions such as the British Isles, the sea breeze is most likely to be felt on quiet, sunny days in summer. It starts late in the morning, is strongest in the afternoon, when it may reach about 10 m.p.h., and it dies down in the evening. It is often felt at places several miles inland. In equatorial regions, where the sun's heat is stronger and other surface winds are relatively light, the sea breeze may reach speeds of 20 to 25 m.p.h.

Land breeze. At night the land loses much heat by radiation and cools more than the sea. The atmospheric pressure over the land is greater than that over the sea, and air moves from the land towards the sea: this moving air is called a *land breeze*. It usually starts about midnight or a little later and dies down after sunrise. Its speed is generally less than that of the sea breeze and it extends a shorter distance beyond the coastline.

NEW VISUAL GEOGRAPHY

The Weather

NEW VISUAL GEOGRAPHY

W. G. Moore

Regional

ICE-CAP AND TUNDRA

THE NORTHERN FORESTS

THE TEMPERATE GRASSLANDS

DESERTS OF THE WORLD

Economic

THE MINING OF COAL

IRON AND STEEL PRODUCTION

THE PRODUCTION OF OIL

Physical

RIVERS AND THEIR WORK

THE SEA AND THE COAST

THE WEATHER

NEW VISUAL GEOGRAPHY

Physical Series

The Weather

W. G. MOORE

HUTCHINSON EDUCATIONAL

HUTCHINSON EDUCATIONAL LTD

178–202 Great Portland Street, London W1

London Melbourne Sydney
Auckland Bombay Toronto
Johannesburg New York

First published 1968

*This book has been set in Baskerville, printed in Great
Britain on coated paper by Anchor Press, and bound
by Wm. Brendon, both of Tiptree, Essex*

09 086420 4 (cased)

09 086421 2 (paper)

Contents

Do You Remember?

Further Work

Acknowledgement is due to the following for permission to reproduce illustrations: Aerofilms Ltd.: No. 8; Balfour Stewart Auroral Laboratory, Edinburgh: No. 23; B.E.A.: Nos. 1, 12; B.O.A.C.: Nos. 9, 11; British Antarctic Survey: No. 15; J. Allan Cash: Nos. 2, 5, 7, 10, 14, 17, 19, 20, 28; Department of Transport, Ottawa: No. 22; A. Eggenberger, Walzenhausen: No. 18; Fox Photos Ltd.: No. 3; Ernest McGill, Wichita Falls: No. 24; Radio Times Hulton Picture Library: No. 27; Royal Meteorological Society: No. 16; Swiss National Tourist Office: Nos. 6, 13; U.A.R. Tourist and Information Centre: No. 4; U.S. Department of Commerce, Weather Bureau: No. 26; U.S. Navy: No. 25; Vannucci Foto-Services, Williamsport: No. 21.

THE WEATHER CHART

The *weather chart* is a map or chart which shows the weather over a large area at a certain time on a certain day. Sometimes it is called a *synoptic chart*, because it gives a synopsis, or summary, of the weather. The weather forecaster, who draws the chart, obtains his information from reports from weather stations in the area. When all these reports have been plotted on the chart he draws the *isobars*, which are lines joining places of equal atmospheric pressure. The pressure is measured in millibars, and the forecaster draws the isobars at intervals of 2 or 4 millibars (see end-papers Nos. 1 and 2). He then divides the chart into regions of high pressure and low pressure.

Anticyclone End-paper No. 1 shows a region of high pressure, or *anticyclone*, over the British Isles and W. Europe. The area of highest pressure is inside the central isobar (1032 mb.). Arrows show the direction of the wind at various places. Notice that the air is circulating clockwise round the anticyclone; in the southern hemisphere the circulation would be counter-clockwise. But the winds near the centre of the anticyclone are very light, and many places reported a calm. Winds are always light when the isobars are far apart.

Depression End-paper No. 2 shows a region of low pressure, or *depression*. The central isobar now encloses the area of lowest pressure. Notice that the winds are blowing counter-clockwise round the depression; in the southern hemisphere the circulation would be clockwise. The winds on this chart were much stronger than on No. 1, especially S.W. of the British Isles: see how close together the isobars are here. W.F. on the chart is a *warm front*. This is a line on the earth's surface along which warm air is rising above cooler air, giving much cloud and rain. C.F. is a *cold front*, a line along which cold air is pushing beneath the warm air. See end-paper diagram No. 3. The area between W.F. and C.F. is called the *warm sector*. The cold front moves more quickly than the warm front. When it overtakes the warm front the resulting front is called an occluded front, or *occlusion*. This is shown by Oc. on the chart.

1 *Weather instruments: measuring the air temperature*

To study the weather properly you must have measuring instruments. One of the most important things to measure is the temperature of the air, and for this a *thermometer* is used. Every weather station has a number of thermometers, which are kept in a wooden box called a *Stevenson screen*. As you see in the picture, the box is painted white to reflect the sun's rays, and the sides are made so that air will flow through easily. The screen is fixed on legs about four feet above the ground.

The upright thermometer on the left is the one which tells the temperature of the air. Look at the other upright thermometer. Its bulb is covered with muslin and kept moist by a wick which dips into a small glass bottle containing water. These two thermometers are called the *dry bulb thermometer* and the *wet bulb thermometer*, and the two together are known as a *hygrometer*. They indicate how moist the air is: there is a big difference between their readings when the air is dry, and a small difference when it is damp. From the two readings the *relative humidity* can be worked out: this is the ratio between the amount of water vapour in a given volume of air and the amount necessary to saturate it at the same temperature, and is expressed as a percentage.

The thermometer that the weather observer in the picture is holding is a *minimum thermometer*, which tells him the lowest temperature that has been reached during the last 24 hours. On the horizontal rack, just above the one where the minimum thermometer was resting, is the *maximum thermometer*. This tells him the highest temperature reached during the last 24 hours. The two other instruments in the screen are the *thermograph* (left) and the *hygrograph* (right). Each of them records continuously by means of a pen which rests on a chart fixed to a rotating drum. The thermograph records the temperature, and the hygrograph records the relative humidity. Can you see the traces that the pens have made? The temperature is rising and the relative humidity is falling.

ANSWER THESE QUESTIONS

1 How high is the Stevenson screen above the ground?
2 Which thermometers measure (a) the lowest, (b) the highest temperatures each day?

2 *Measuring wind direction: smoke rising vertically*

One of the most important weather instruments is the *barometer*, which gives the atmospheric pressure. A self-recording barometer is called a *barograph*. Another instrument is the *rain gauge*, which indicates the depth of rain that falls at a place. The *wind vane*, or weather vane, gives the direction of the wind.

You can find out a great deal about the weather without any instruments. Look at the smoke coming from the Yorkshire colliery chimney in the picture: it is rising vertically into the air. This indicates that here the air near the earth's surface is calm: there is apparently no wind. Such a calm often occurs in an anticyclone. On end-paper diagram No. 1 many weather stations in southern England and northern France, near the centre of the anticyclone, reported a calm. When there is a wind its direction is described by the point of the compass from which it blows. If the chimney smoke began to drift slowly towards the north we should say that a 'light southerly breeze' had sprung up.

The weather changes a great deal with the direction of the wind. In Great Britain, for instance, a northerly wind usually means cool or cold weather; a southerly wind brings mild or warm weather. You can judge the direction of the wind anywhere by studying the weather chart. Look at the arrows on end-paper diagrams Nos. 1 and 2. They lie at a slight angle to the isobars, and you can determine their direction by this simple method: if, in the northern hemisphere, you imagine yourself standing with your back to the wind, then the low pressure is always on your left. In the southern hemisphere the low pressure is on your right.

If the wind changes in a clockwise direction, from S. to S.W. and then W., it is said to be *veering*. If it changes in a counter-clockwise direction, for example from W. to S.W. and then S., it is said to be *backing*. Suppose that the depression in end-paper diagram No. 2 is moving due E. The wind will gradually veer over England and Wales as the depression crosses the country.

ANSWER THESE QUESTIONS
1 With what instrument can you measure rainfall?
2 If the wind changes from S. to S.E. is it veering or backing?

3 *Measuring wind speed: gale in the English Channel*

Wind is simply air in movement. When the air moves with great speed we say that the wind is strong. If the wind is very strong indeed we say that there is a gale. The picture shows a gale in the English Channel, taken on a ferry steamer between Southampton and the Isle of Wight. Huge waves are pounding against the bows, and the high wind is carrying spray right over the cyclists and the cars.

In English-speaking countries the speed of the wind is measured in miles per hour. In the gale in the picture the wind reached a speed of more than 60 m.p.h. and was much stronger than in most gales. How can you estimate the wind speed? One way is to use the *Beaufort scale*, which was invented by Admiral Beaufort in the early nineteenth century. This scale ranges from 0 to 12. 0 represents a calm, which is shown by smoke rising vertically (see No. 2); 1 is shown by the slow drift of smoke from a chimney (1 – 3 m.p.h.), and 2 by the leaves of trees rustling (4 – 7 m.p.h.). The gale in the picture was strong enough to uproot trees, equal to force 10 on the Beaufort scale (55 – 63 m.p.h.). Force 7 on the Beaufort scale represents a moderate gale and is shown by whole trees being moved by the wind (32 – 38 m.p.h.). Evidently the gale in the picture was one of unusual strength. Gales are shown on a weather chart by the closeness of the isobars: on end-paper No 2, for instance, there were gales in the Bay of Biscay.

Many weather stations have instruments which measure wind speeds much more accurately than can the Beaufort scale. These instruments are called *anemometers*. The main types of anemometer comprise three or four metal cups mounted on a vertical spindle. These cups rotate as the wind blows, and their rate of rotation gives the speed of the wind. In one type of anemometer the wind speed can be read on one dial and the wind direction, taken from a wind vane, on a second. A self-recording anemometer is called an *anemograph*.

ANSWER THESE QUESTIONS

1 What is the Beaufort scale of wind speeds?
2 What is the instrument which measures wind speed? What is the self-recording type called?

4 *High sun: in Port Fuad, United Arab Republic*

You cannot actually see the sun in the picture, but this is what interests us here. The photograph was taken in summer in Port Fuad, near Port Said, in the United Arab Republic (Egypt). The clock on the tower in the background indicates that the time was almost 2 p.m. This is the hottest, brightest time of the day: the cyclists are wearing dark spectacles to protect their eyes from the glare of the sun. See how short their shadows are. This means that the sun is very high in the sky. At the summer solstice (about 22nd June) the sun is less than 8° below the vertical at noon. Is the sun ever as high as this where you live? It is never very high in the sky at places of high latitude: temperatures in high latitudes are generally lower than they are in low latitudes. The mean annual temperature at Port Fuad is 68°F. (20°C.), compared with 50°F. (10°C.) at London, England.

Besides having higher temperatures than countries in higher latitudes the U.A.R. also has a much greater *duration of sunshine*. This is the number of hours of bright sunshine in a given period, for example a year. At many weather stations in different parts of the world the duration of sunshine is measured with an instrument called a *sunshine recorder*. This consists of a spherical glass lens which focuses the sun's rays on a card marked off in hours. The rays burn a line across the card, recording the number of hours that the sun has been shining on any day.

The sun is the only source of heat for the earth. Wherever you live it is the sun that controls the weather. The radiant energy that reaches the earth from the sun, travelling an average distance of 93 million miles through space, is called *insolation*. The amount of insolation varies a great deal in different parts of the earth's surface: at the equator and in low latitudes the insolation is much greater than in high latitudes.

ANSWER THESE QUESTIONS

1 The mean annual temperatures at Stockholm, Berlin, and Rome are 5·6°C., 9·1°C., and 15·4°C. respectively. What are their latitudes? (Use your atlas.)

2 How does a sunshine recorder tell the duration of sunshine?

5 Sunset: on the coast of Scotland

The amount of insolation at any place varies with the angle of the sun. The insolation is greatest when the sun is at its highest point in the sky, at noon. Some time is needed for the sun's rays to warm the earth's surface, and then for the earth to heat the atmosphere. Thus the maximum air temperature is usually reached some time after midday, in the early afternoon.

As the sun sinks and its rays become more oblique, the insolation decreases. The time that the sun reaches the horizon is called the *sunset*. Strictly speaking, it is the time when the centre of the sun touches the horizon. As you can see, the picture was taken just before sunset. Notice the very oblique rays reflected in the water; these have to pass through a much greater thickness of the atmosphere than more direct rays. The rays of shorter wave length in the sunlight (chiefly blue) are scattered by the gases and dust particles of the atmosphere, but those of greater wave length (yellow, orange, red) pass through. This is why sunsets often give a display of brilliant yellow, orange and red. Their beauty is even greater when, as in the photograph, broken clouds are lit up by these colours. Insolation is cut off when the sun sinks below the horizon, and the air temperature falls, sometimes rapidly. The minimum air temperature is often reached just before *sunrise*—the time when the centre of the rising sun is just on the horizon.

Why does the darkness not come suddenly, just after sunset, and continue until sunrise? When the sun is not far below the horizon its light is reflected down to the earth's surface from the upper layers of the atmosphere. This faint reflected light before sunrise and after sunset is called the *twilight*. *Civil twilight* is the light we receive between the time that the sun is 6° below the horizon and sunrise, or between sunset and the time that it is 6° below the horizon. It is still possible to do outdoor work in this light. *Astronomical twilight* lasts from and to the time when the sun is 18° below the horizon, and is understood to mark the end (or beginning) of complete darkness.

ANSWER THESE QUESTIONS

1 Why is the sky sometimes red at sunrise or sunset?
2 What is the difference between civil twilight and astronomical twilight?

5

6 *Good visibility: in the Swiss Alps*

Probably your first impression from this picture is of the great distances you can see across the mountain peaks. Look at the climbing party at the opposite end of the ridge in the foreground. Notice how clearly they stand out against the mountain slopes. The photograph was taken on the mountain called the Säntis, which rises to 8,209 feet, in Appenzell Canton, N.E. Switzerland. The line of peaks beyond is the Churfirsten Range. Notice how sharply these peaks are seen against the more distant mountains, yet the nearest of them (left) is 7 miles from the Säntis. The mountains in the distance (across the top of the picture) are still very clearly visible, although they are about 25 miles away. The most distant peaks are about 50 miles away.

The greatest distance that you can see in a horizontal direction is called the *visibility*. It is a very important item in weather reports, especially when they are being prepared for aircraft. Since the observer has no instruments to help him find the visibility, he has to estimate the distance. First he selects a number of prominent objects at known distances from the weather station. The visibility is measured by the distance of the farthest object that he can clearly see in daylight. After dark he generally makes use of lights at different distances.

Visibility depends on the number of solid or liquid particles in the atmosphere. When there is a great quantity of solid particles in the air, for example dust or smoke, the visibility is poor, for light cannot pass through these particles. Visibility is also made poor by large numbers of water particles—rain, snow, mist, or fog. In a mountainous region like the one in the picture the air contains very few solid or liquid particles and the visibility is therefore good. When flying at a great height people usually find that the visibility in a horizontal direction is far better than it is on the earth's surface. It is sometimes possible to recognise a mountain peak at a distance of well over 100 miles.

ANSWER THESE QUESTIONS

1   How would you estimate the visibility at night?
2   Why is the visibility usually better in the mountains than at lower altitudes?

7 *Poor visibility: a mist over the River Thames*

In this picture, as in No. 6, the sun is shining (notice the shadows), but here the visibility is poor. This is because the air contains a large number of water droplets—a condition of the atmosphere which we describe as a *mist*. How and why does a mist form? You must remember first that there is always a certain amount of water vapour in the atmosphere, even in dry weather. Suppose that the air near the earth's surface is cooled, say under a clear sky at night, when the earth loses much heat by radiation. Its temperature may fall so much that the water vapour is condensed, and a mass of tiny water droplets is formed. This may result in either a mist or a fog. If the visibility is reduced below 2,000 metres (2,200 yards) but remains above 1,000 metres (1,100 yards), we say that a mist has formed. If the visibility falls below 1,000 metres, we use the term *fog*.

In the picture we are looking down the River Thames from Billingsgate, London. Notice how much less clear Tower Bridge appears than did the mountain peaks in No. 6. Yet the bridge is less than half a mile away. The cranes on each bank beyond the bridge and the boat seen through the arch of the bridge are only just visible. They are about $\frac{3}{4}$ mile away, and the visibility is therefore between 1,100 and 2,200 yards. In mountainous or hilly regions a low cloud may cover the ground in places, resembling a mixture of mist and drizzle. This is often called a *Scotch mist,* because it is common in mountainous Scotland. In some areas it is known as *mizzle* (mist plus drizzle).

If the poor visibility is caused by particles of smoke or dust, instead of by water particles, the term *haze* is used in place of mist. Smoke enters the atmosphere from factory and house chimneys, and haze is much commoner in industrial areas than in rural districts. With a light wind, however, the smoke sometimes drifts many miles from its source. Dust is often raised by the wind from the surface of a desert, and again it may be carried over great distances (see No. 28).

ANSWER THESE QUESTIONS

1 What is the difference between mist and fog?
2 Why is haze commoner in big towns than in the open country?

8 Bad visibility: fog in Hertfordshire, England

In the place where this picture was taken the visibility on the ground was much worse than in No. 7. The photograph was taken from an aircraft flying over the fog in Hertfordshire, England. From above, the fog looks just like a uniform layer of cloud—a type of cloud known as *stratus*. In fact, if such fog occurs on hills, it may be reported either as low stratus cloud, or as *hill fog*. Notice that the fog is so shallow that the tops of the factory chimneys project above it. Fog is usually less than 1,000 feet in depth, and often less than 500 feet—as it is in the picture. At sea the fog is sometimes so shallow that it does not reach the top of a ship's mast. Can you see the darker areas towards the right of the picture? In these spots the fog is a little thinner and the visibility is better than elsewhere. Weather observers describe such fog as 'patchy'. You will remember that a fog is reported when the visibility is less than 1,000 metres (1,100 yards). It is said to be a 'thick fog' when the visibility is less than 200 metres (220 yards).

Fog, like mist, is formed when the air is cooled and condensation takes place. The air may be cooled at night, for instance, when the land loses heat by radiation. A light wind will spread the cooling effect through a certain depth of air, and a fog will form to that depth: this is known as a *radiation fog*. Sometimes warm, moist air moves across the cold surface of the land or the sea and is chilled. Condensation takes place and an *advection fog* is formed. (Advection means movement in a horizontal direction.) The water droplets condense on minute solid particles floating about in the atmosphere. In a busy industrial area there is often a large quantity of smoke particles in the air and the water droplets condense on these. We call this kind of fog a smoke fog or *smog*.

In winter, radiation fog or mist often occurs in an anticyclone. On the chart of end-paper No. 1 several places in southern England and northern France reported fog or mist.

ANSWER THESE QUESTIONS

1 What is the usual depth of a fog?
2 How are radiation fogs and advection fogs formed?

9 *Low cloud: stratocumulus over the Pacific Ocean*

The aircraft in this picture is flying at a much higher level than in No. 8. It is high above the sheet of cloud which covers much of the picture. The cloud itself is about 2,000 feet above the surface of the sea; the picture was taken over the Pacific Ocean near Hawaii. You can see the ocean through breaks in the cloud (for example, bottom right). Notice that the cloud is not very deep but that it extends over a great area; this is known as a *layer cloud*. This cloud looks different from the stratus cloud mentioned in No. 8, which is also a kind of layer cloud. Its surface is not flat and uniform, and it consists of long rolls and small rounded masses: it is called *stratocumulus* cloud. The lower surface has much the same wavy appearance as the upper surface. Sometimes this type of cloud covers the whole sky, but it may be broken, as in the picture, with patches of blue sky showing between the pieces of cloud.

Clouds, like fog, are formed when part of the air is cooled, leading to the condensation of water vapour. The difference between them is that clouds form above the earth's surface, whereas fog forms at ground level. Stratocumulus cloud is often formed when the lower air is churned up within a fairly shallow layer by the wind—a process known as *turbulence*. Some of the air is cooled, condensation takes place, and cloud is formed. Such cloud does not usually mean wet weather— at worst some drizzle or light snow. If the turbulence takes place at much higher levels in the atmosphere the resulting cloud is called *altocumulus*. A similar cloud formed at the highest levels is called *cirrocumulus*. Both these types of cloud look much like stratocumulus, but their layers are generally thinner.

ANSWER THESE QUESTIONS

1 What process in the atmosphere leads to the formation of stratocumulus cloud? How does this process take place?
2 What are the differences between stratocumulus, altocumulus, and cirrocumulus clouds?

10 *Low cloud: fair-weather cumulus over the prairie*

In this picture, taken on the prairie in Alberta, Canada, you are looking at the clouds from the ground. As you can see, the clouds are deeper than those in No. 9. All clouds which extend upwards to a considerable depth are known as *heap clouds*, and this one is called a *cumulus* cloud. You often see these puffs of white cloud drifting across the sky on warm summer days. Because they form during spells of fair weather, and do not give rain, they are called *fair-weather cumulus*.

Such clouds are formed by convection currents in the atmosphere. Convection means upward movement. Advection, you may remember, means horizontal movement (see No. 8). The sun warms the earth, which in turn heats the air just above it. This heated air rises in a convection current and eventually reaches a level at which it has been so much cooled that condensation takes place. Fair-weather cumulus clouds form in the late morning and early afternoon because that is when the sun's heat is strongest and causes the greatest convection. They disappear in the evening when the sun is no longer heating the ground and convection currents have ceased. When the air rises to a great height above the earth's surface, a cloud of very considerable depth forms (see No. 11). Look at the lower surfaces of the clouds in the picture. Can you see that they are all at about the same level above the ground? This was the level at which condensation began as the air was rising—the *condensation level*.

If you were standing on the prairie with the two horsemen do you think you could estimate how much of the sky is covered with cloud? This is what a weather observer has to do. He estimates what fraction (in eighths) of the sky is covered with cloud, and sends this information as part of his weather report. He also estimates the height of the cloud base above the ground and includes this in the report.

ANSWER THESE QUESTIONS

1 How do heap clouds differ from layer clouds?

2 If you were a weather observer how would you report the amount of sky covered with cloud?

11 *Clouds at different heights: low, medium, high*

Clouds are classified according to the heights of their bases above the ground as low, medium, or high clouds. The clouds in Nos. 9 and 10 were low clouds, for in both cases the base of the cloud was not very far above the ground. This picture shows clouds of all three heights and also clouds of both layer and heap types. Can you identify the cloud in the foreground, stretching far into the middle distance (right)? It is a layer of stratocumulus cloud, with small cumulus clouds pushing up through it here and there. On the left of the picture is a massive cumulus cloud, towering above the stratocumulus layer. It is still called a low cloud although its top is so high, for its base is at about the same level as the stratocumulus cloud. It develops to these enormous depths—several thousand feet—when strong convection currents force air upwards, and cauliflower-shaped tops like those in the picture (upper left) are formed. All clouds with a base between the earth's surface and about 6,500 feet are classified as low clouds, and they include the stratus and stratocumulus (layer) and cumulus (heap) types.

Look at the cloud in the middle right of the picture. It is flat and uniform like stratus, but much higher than the stratocumulus: it is a medium cloud called *altostratus*. Such medium clouds occur at heights between about 6,500 and 20,000 feet. If the cloud at such a height is a heap rather than a layer cloud, consisting of small rounded masses with blue sky between them, it is called altocumulus cloud. A popular name for a sky covered with this cloud is '*mackerel sky*', because it looks something like the markings on a mackerel. Now look at the small feathery tufts of cloud above the altostratus cloud. This is a high cloud, above 20,000 feet, and is called *cirrus* cloud. If such a high cloud resembles stratus, it is called *cirrostratus* cloud; if it is more like cumulus, it is called cirrocumulus cloud.

ANSWER THESE QUESTIONS

1   At what heights above the earth's surface are low, medium and high clouds to be found?
2   What are the names of the two kinds of medium clouds? How would you describe them?

12 *Cumulonimbus cloud and shower: over Renfrewshire, Scotland*

One of the things we noticed about the clouds in picture No. 10 was the uniform level of their bases, and you can see the same thing in this picture. Look at the deep cloud in the centre of the picture. This is a *cumulonimbus* cloud, which is also classified as a low cloud because its base is not very high above the earth's surface. It develops from heavy cumulus cloud (see No. 11) when the rising air continues to even greater heights. Eventually the air becomes so cold that water vapour changes to ice crystals instead of water droplets, as it does when cirrus clouds are formed. As the picture shows, the top of a cumulonimbus cloud often spreads out into the shape of an anvil, and parts of it look like cirrus (right). The cloud has great depth, perhaps up to 30,000 feet in temperate regions and much higher in the tropics. From an aircraft it shines in the sunlight, but from beneath it is dark and threatening, and is sometimes associated with thunderstorms (see No. 14).

A cumulonimbus or heavy cumulus cloud often produces a *shower* of rain, which may consist of large drops but only lasts for a short time. Ordinary rain or drizzle, whether it continues for a long time or not, consists of much smaller drops and falls from layer clouds such as altostratus or *nimbostratus*. Look at the space beneath the clouds on the left. The visibility is good, for the distant hills can be clearly seen. But from the centre of the cumulonimbus cloud a shower of rain is falling, and the hills cannot be seen: whenever rain falls the visibility becomes poorer. *Hail* sometimes falls from a cumulonimbus cloud, even in summer. Hail consists of pellets of ice, perhaps only a few millimetres in diameter but often much larger. Hailstones as big as marbles have frequently fallen in Britain, and elsewhere they have reached the size of billiard balls or even oranges. Hail is probably formed when water droplets are carried by violent air currents into a region where the temperature is below the freezing point. The droplets freeze, and the hailstones grow larger and larger as the further layers of ice freeze on to them.

ANSWER THESE QUESTIONS

1 How does a cumulonimbus cloud form?
2 What is hail? How do hailstones form and how do they increase in size before they fall?

13 *Banner cloud: on the Matterhorn, Switzerland*

The high mountain in the upper part of the picture is the Matterhorn (14,780 feet), in Switzerland. To the left of the mountain is a large cloud streaming out from the summit like a banner or flag. It is a special kind of cloud seen at times on the peaks of a number of mountains and is known as a *banner cloud*. It is sometimes seen, for example, at Gibraltar. There it extends westwards from the summit of the Rock when a fairly strong east wind is blowing. The banner cloud does not move across the sky like other clouds, but seems to be attached to the mountain or rock.

The usual explanation for the formation of a banner cloud is that a strong wind forces air up the *windward* slope of the mountain—the slope facing the wind. This is on the right side of the mountain in the picture. The rise of air causes cooling, and the water vapour that it contains condenses to form the cloud on the *leeward* side of the mountain—the side sheltered from the wind (on the left of the mountain in the picture). Some distance down wind the air descends, which causes it to be warmed again. The water droplets of the cloud evaporate and so the cloud ends there. Although there is a constant stream of air through the banner cloud, the cloud itself stays in the same position, apparently attached to the mountain summit.

ANSWER THESE QUESTIONS

1 How is a banner cloud formed? Why does it seem to remain stationary when a strong wind is blowing?
2 Does a banner cloud stream out from a mountain summit on the windward or the leeward side?

14 *Lightning flashes: cumulonimbus cloud and thunderstorm*

A *thunderstorm* is a storm which takes place with a cumulonimbus cloud and includes lightning and thunder. Often there is a heavy shower with the storm, and sometimes hail as well as rain. The flashes of *lightning* in the thunderstorm in the picture are discharges of electricity, which are caused by the breaking up of large raindrops by the violent air currents in the cloud. This results in an enormous positive charge building up in the top of the cloud, and a vast negative charge below. The charges become so large that at last an electrical discharge takes place within the cloud: the lightning flashes on the left and right of the picture are of this kind. Sometimes the discharge takes place from cloud to ground, as in the central flash. This is the dangerous kind of flash, which may cause damage and even loss of life. When the flashes are visible, as they are in the picture, they are known as *forked lightning*. *Sheet lightning* is the light of distant flashes seen through the clouds. The noise of thunder is caused by the sudden expansion of the air due to the heat of the lightning flash, followed by contraction.

Thunderstorms are most frequent in warm summer weather, when the heating which causes convection currents and the formation of cumulonimbus clouds is greatest. They may also occur along a cold front (see p. vii), owing to warm air rising vigorously upwards over cold air, even in winter. The frequency of thunderstorms varies with location. In the British Isles, for instance, they occur most often in east and south-east England (15 to 20 times a year), where summer temperatures are highest. Thunderstorms are most frequent near the equator: places in the islands of Indonesia and the Amazon basin have as many as 200 or more during the year.

ANSWER THESE QUESTIONS

1 What are the causes of lightning and thunder in a thunderstorm?
2 In what part of the world are thunderstorms most frequent? Where are they most frequent in the British Isles?

15 *Cirrostratus cloud, halo and mock sun: in Graham Land, Antarctica*

The bright light on the right of the picture, which was taken in the Hope Bay district, Graham Land (British Antarctic Territory), is the sun. We cannot see it clearly because it is shining through a thin sheet of cirrostratus cloud. When the sun shines through the ice crystals of the cloud the light is bent, or refracted, causing a ring of light, or *halo*, to appear. Usually the halo is white, but if it is very well developed the inner edge is red. Part of the halo may be seen in the picture (left). The sun is at the centre of the circle formed by the halo, and the angle between sun and halo is $22°$. A similar halo may be formed when the moon shines through cirrostratus cloud. The halo is believed by many people to be a sign of bad weather, probably because cirrostratus cloud and haloes are sometimes seen ahead of a warm front. It occurs too commonly to be a reliable guide, however.

Can you see the faint band of white light passing across the picture through the centre of the sun? This is part of the *parhelic circle*, which is parallel to the horizon and is caused by the reflection of sunlight through the ice crystals of the cloud. Just above the roof of the hut (left), at the point where the parhelic circle and the halo intersect, is a blob of light, like a small image of the sun. It is a *mock sun*, or *parhelion*, and is due to the refraction of sunlight through the ice crystals. When the sun is near the horizon, as it is in the picture, the parhelion is very bright, and is red on the side nearer to the sun and blue on the far side. A similar image of the moon, which is less bright, is called a *mock moon* or *paraselene*.

A circular band of light round the sun or moon that is much smaller than the halo is called a *corona*. It consists of several coloured rings, reddish on the outside and bluish on the inside, and is formed by the diffraction (another type of bending) of light by the water droplets of a cloud.

ANSWER THESE QUESTIONS

1 How is a halo formed round the sun or moon?
2 Where is a mock sun seen when the sun is near the horizon?

16 *Sunlight and shower: the rainbow*

The picture shows part of a *rainbow*, a familiar sight when the weather consists of 'sunny intervals and showers'. Its shape is the arc of a circle, and it is seen when sunlight falls on rain. Notice in the photograph that the rainbow is in front of the camera while the sun is behind it: the sun, the camera (or the observer's eye) and the centre of the arc are in the same straight line. This is what happens when light from the sun falls on a single drop of rain: as the light enters the drop it is refracted, or bent; it is then reflected from the far side, and refracted again as it comes out of the drop. When the white light is refracted it is also split up into the different colours of the spectrum. Thus the rainbow is seen not as an arc of white light but as a band of colours, with red on the outside and violet on the inside. The appearance of the colours depends on the size of the rain-drops. When the drops are only of medium size the colours are much less bright and distinct than when the drops are large. The water drops in a fog are very small indeed, and the special kind of bow sometimes seen in a fog and known as a *fog bow* is practically white.

Look carefully above and to the right of the rainbow in the picture. Can you see a second, much fainter bow outside the first? The inside bright rainbow is called the 'primary bow' and the outside fainter one is called the 'secondary bow'. This 'secondary bow' results from the white light being reflected twice inside each raindrop, and its colours are therefore in opposite order to those of the 'primary bow': red is on the inside and violet on the outside of the band. Notice that the two bows are concentric—they have the same centre. The angle subtended by the radius (the angular radius) of the 'primary bow' at the observer's eye is about $42°$, while the angular radius of the 'secondary bow' is about $54°$.

ANSWER THESE QUESTIONS

1 Why are two rainbows sometimes seen at the same time? How do their colours differ?

2 What is a fog bow?

17 *Snow falling: at Svolvaer, Lofoten Islands, Norway*

The picture shows a fall of *snow* at Svolvaer, the chief town in the Lofoten Islands, in Norway. Perhaps you can see some of the *snowflakes* on their way down to the ground. These snowflakes are collections of minute ice crystals. They are generally hexagonal (six-sided), but probably no two snowflakes are ever exactly alike. Snow forms instead of rain when water vapour condenses at temperatures below *freezing point* (0°C. or 32°F.). If the temperature is very low the snow forms as small grains called granular snow. The presence of the many snowflakes in the atmosphere lowers the visibility, which in a heavy fall of snow may be only a few yards. If a strong wind carries the falling snowflakes and perhaps lifts snow from the ground, the storm is called a *blizzard*. The visibility in a blizzard may be reduced almost to zero. Even in the light snow shower in the picture the buildings across the harbour are much less clearly visible than they would be on a fine day.

An easy method of measuring snow is to find its depth on the ground with a ruler. A more usual way is to take the snow that has collected on the rain gauge, then to melt it carefully and measure it in the graduated cylinder used for measuring rain. The *snow gauge* is generally only a rain gauge fitted with a shield to help collection and accurate measurement. All forms of *precipitation*, or moisture deposited on the earth's surface, whether rain, snow, hail, or any other, are expressed as a depth of *rainfall*, in inches or millimetres. The amount of water in snow varies greatly with the type of snow, but in general 10 or 12 inches of snow are reckoned as equivalent to 1 inch of rain.

When the picture was taken, the air near the earth's surface was so cold that none of the snow was melting. If snowflakes descend through a layer of warm air, however, they reach the ground as rain. In Great Britain falling snow that is partly melted, or a mixture of snow and rain, is called *sleet*.

ANSWER THESE QUESTIONS

1 How is snow usually measured? How much snow is equivalent to 1 inch of rain?
2 What are (a) granular snow, (b) sleet, (c) a blizzard?

17

18 *Snowdrift in the mountains: the Churfirsten Range, Switzerland*

In general, places situated in high latitudes have more snow than places in low latitudes. In the Scilly Isles, off the south-west coast of England (about latitude 50°N), snow falls on only three or four days a year. At Svolvaer, Lofoten Islands (No. 17), off the north-west coast of Norway (about latitude 68°N.), snow falls at least ten times more often. Both frequency and amounts of snowfall increase with altitude. In the Scilly Isles (near sea level) snow scarcely ever collects to any measurable depth. On the Säntis (No. 6), in a lower latitude (about 47°N.) but at 8,209 feet above mean sea level, over 40 feet of snow collects in the winter.

This picture was taken near the Säntis and the Churfirsten Range, in Switzerland. It shows both the depth of snow at high altitude and the formation of a *snowdrift*. When the snow was falling here, a strong wind was blowing. This carried much of the snow to the leeward or sheltered side of the hut, in the foreground of the picture, where it collected in a deep bank, or snowdrift. In hilly or mountainous districts such snowdrifts form on the sheltered sides of walls and hedgerows and may bury sheep and even human beings. They also form in sheltered hollows on mountain slopes and sometimes remain into the summer, long after the surrounding snow has melted.

When snow is no longer falling, an observer reports 'snow lying' if half or more of the land around his weather station is still snow-covered. In mountainous districts, such as the one in the picture, there will be reports of 'snow lying' for many more days than in low-lying areas. If the snow is very deep, the danger of an *avalanche* may arise. An avalanche, a gigantic mass of snow and ice and possibly rock sliding down a mountain slope, may cause vast destruction to villages and roads. It often sets up a strong wind known as an *avalanche wind*, which may cause damage at places far from the avalanche itself.

ANSWER THESE QUESTIONS

1   How does snowfall vary with latitude and altitude?

2   How and where do snowdrifts form?

19 *Condensation near the ground: dew on a spider's web*

Look at the tiny drops of water on the strands of the spider's web in the picture. They consist of *dew*, formed when water vapour in the air near the ground condensed. At what time of day have you seen dew in the garden or elsewhere? It forms during the night, so you have probably observed it in the early morning. On a warm, damp day there is a great deal of water vapour in the lower layers of the atmosphere. If this is followed by a night of clear skies the earth will lose heat by radiation, and the air in contact with it will be cooled. If the air is cooled enough its water vapour will condense on the ground and on objects near the ground, such as blades of grass, or leaves, or a spider's web.

In order that the air shall be long enough in contact with the ground for condensation to take place there must be a calm or only a very light wind. Under these conditions the air nearest the ground will be cooler than the air above. This is the opposite of the usual decrease of temperature with height above the earth's surface and is called an *inversion of temperature*, or temperature inversion. If there is much movement in the lower air, that is, a surface wind, the air close to the ground will not be sufficiently cooled for condensation to take place.

The temperature to which the air must be cooled in order that its water vapour shall condense is called the *dew point*. At this temperature the air is saturated with water vapour, so that further cooling leads to condensation in the form of dew. The dew point can be worked out by reading the wet and dry bulb thermometers or hygrometer (No. 1). It is plotted for each station on the weather chart, along with the air temperature, and helps the forecaster to predict fog, frost, or dew, and the height of the base of cumulus clouds (No. 10).

ANSWER THESE QUESTIONS

1 In what weather conditions does dew form on and near the ground?
2 What is the dew point? What instrument is used to work it out?

Ice crystals near the ground: hoar frost on plants

The white deposits on the plants in the picture may look very different from the dew in No. 19, but they were formed in the same way from water vapour. They consist of ice crystals, like tiny needles, and are known as *hoar frost*. Perhaps you have looked out on a clear winter morning to see lawns and pavements 'white with frost'—just as you may have seen them wet with dew on a bright summer morning. Hoar frost forms instead of dew when the dew point of the air is below the freezing point (32°F. or 0°C.). When the air at this low temperature becomes saturated with water vapour, the water vapour does not condense to drops of water (dew), but changes directly to the solid state (ice). The weather conditions that lead to the formation of hoar frost at night are similar to those for dew: the sky must be clear and there must be little or no wind. In the early part of the night the dew point may be above the freezing point, in which case dew will form. If the air temperature then falls below the freezing point, the dew will freeze. The final deposit on plants and blades of grass will be a mixture of frozen dew and hoar frost.

Another kind of ice crystal deposit which looks something like hoar frost but is formed in a different way, is known as *rime*. You might describe it very briefly as 'frozen fog'. If the temperature in a fog is below the freezing point the water droplets in the fog are said to be supercooled: they are still in the liquid state although below the temperature at which they normally freeze. When a wind brings these supercooled droplets into contact with a solid object at a temperature below 0°C. they immediately freeze on it as minute ice crystals. So a deposit of rime forms on the windward sides of fences, poles, etc. On mountain slopes the supercooled water droplets of a low cloud may also cause rime. Such deposits, sometimes called 'frost feathers', have been known to grow out to a distance of several feet.

ANSWER THESE QUESTIONS

1 When does hoar frost form instead of dew?
2 How is rime formed?

21 *Damage by ice: glazed frost on overhead wires*

This picture from the United States shows a linesman holding an overhead wire which has been brought down by a thick deposit of ice. The deposit is called *glazed frost* or *glaze* and consists of frozen rain. In the picture the linesman is starting his repair work by knocking the ice from the wire. Notice that the deposit is almost as thick as his arm. How does such a mass of ice collect on the wire? In the first place the weather must be very cold, and the surfaces of objects near the earth's surface must be below freezing point. If a warm upper layer of air then reaches the area, with the approach of a warm front, for instance, rain may fall on the cold surfaces. The rain freezes on coming into contact with these surfaces, and the coating of smooth ice may grow to a considerable thickness. If it collects on overhead wires it may become heavy enough to bring them down, and break the branches of trees.

One of the greatest dangers of glazed frost is to aircraft. If a plane is flying through air that is below freezing point while rain is falling, glazed ice may collect on its surface. This may so affect its flight that the pilot has difficulty in keeping control. There is a similar danger if rain falls on roads and footpaths which are at temperatures below 0°C. A thin sheet of ice forms, making it difficult for drivers to control their vehicles and for pedestrians to keep their feet. This kind of glazed frost on roads is sometimes known as *black ice*. It may also occur after a temporary thaw. Suppose a layer of snow on the road is melted by the sun, for instance, and this is followed by a sudden fall of temperature after sunset. The melted snow may become a sheet of ice, and be extremely hazardous to traffic.

ANSWER THESE QUESTIONS

1 What is the difference between glazed frost and hoar frost (No. 20) (a) in the way they are formed, (b) in their appearance?

2 How may black ice form on a road surface? How does it affect traffic?

22 *Effect of frost: ice on the St. Lawrence River, Canada*

The word frost does not always mean something you can see, such as hoar frost (No. 20), or glazed frost (No. 21). Whenever the temperature of the air falls below freezing point (0°C. or 32°F.), we say that there is a *frost*. In some parts of the world the air temperature never falls to such a low level: frost is therefore unknown. In other regions frosts may continue for much of the year. Eastern Canada is such a region, and in the picture you can see an important effect of a long period of severe frost. It was taken on the St. Lawrence River at Montreal, and shows an icebreaker at work on the frozen river.

In November the mean temperature here is just above 0°C., but for the next four months it is well below freezing point. The river freezes, and for most of that period ships cannot use the port. You can see how thick the ice becomes, in the bottom left-hand corner of the picture. In March, when the picture was taken, the icebreakers begin to open up a channel along the river so that navigation may begin again as early as possible.

Every winter many other rivers and lakes in Canada, the U.S.S.R. and elsewhere are frozen for several weeks. In Britain, where winters are so much milder, long periods of severe frost are almost unknown: the River Thames in London, for instance, has been frozen only two or three times during the last 200 years. When the air temperature is only just below freezing point, there is said to be a slight frost. If the temperature is lower, the frost is classified as moderate, severe, or very severe. These different kinds of frost represent the air temperature as measured by the thermometer in the Stevenson screen (see No. 1). Sometimes the air temperature in the screen is just above the freezing point when the temperature at ground level is just below 0°C. This is described as a *ground frost*.

ANSWER THESE QUESTIONS

1 What does a weather forecaster mean when he says that there will be a slight frost at night?
2 What is the difference between a ground frost and an air frost?

23 The northern lights: night sky in Iceland

Many people consider the band of light shown in the picture to be the most beautiful sight in nature. The photograph was taken at Reykjavik, in Iceland, and shows the *aurora borealis* or *northern lights*. People who live in high latitudes, like the inhabitants of Iceland, have a much better chance of seeing this light display than those who live in lower latitudes. In Great Britain, for example, keen observers in southern England might see the aurora on five or six nights during the year. In northern Scotland it could be seen about ten times more frequently. Still farther north, in a belt running north of Norway, south of Iceland, and across northern Canada and the U.S.S.R., it is visible on almost every clear night. This belt, called the auroral zone, passes right round the north magnetic pole of the earth. In high latitudes of the southern hemisphere a similar display known as the *aurora australis* or *southern lights* is seen.

Sometimes the aurora is no more than a quiet glow just above the horizon, without movement or colour. The display in the picture is a bright folded band or arc of light well above the horizon. Its lower edge is clear and sharp, while faint streamers of light flash upwards into the sky from its upper edge. The brightest auroras are wonderful displays of both movement and colour, with yellowish green and red predominating.

What is the aurora? All the facts have not yet been discovered, but it is known that it occurs high in the upper atmosphere, usually about 60 to 70 miles above the earth's surface. It is caused by electrically charged particles from the sun entering the upper atmosphere, and is connected with the appearance of sunspots on the surface of the sun. At times of very active aurora, radio and telegraphic transmission may be seriously affected.

ANSWER THESE QUESTIONS

1 In which region would you be more likely to see the northern lights: (a) in the Shetlands or the Scillies, (b) in Norway or Italy, (c) in Alaska or Florida?

2 What are the names for the aurora of the southern hemisphere?

24 *A tornado, or twister: at Wichita Falls, Texas, U.S.A.*

A *tornado* is a storm which takes the form of a fierce whirlwind. The tornado in the picture is just approaching the town of Wichita Falls, in Texas, U.S.A. Look at the dense black cloud across the top of the picture: this is the base of an enormous cumulonimbus cloud accompanying the tornado, with lightning flashing from the cloud to the ground, thunder claps, and heavy rain. The word tornado comes from the Spanish *tronado,* which means thunderstorm. Notice the funnel-shaped cloud which extends from the cumulonimbus to the ground. As the tornado moves along, the funnel cloud twists and bends, and for this reason a tornado is often known in the United States as a *twister.* When the funnel cloud comes near, the sky darkens and there is a tremendous roaring sound made by the wind and thunder. Buildings are wrecked, trees uprooted, and motor-cars and other heavy objects may be lifted and flung about: during this Wichita Falls tornado over three million dollars' worth of damage was done and seven people were killed.

The tornado usually forms on an active cold front (see p. vii), where warm, moist air is already rising strongly over a wedge of cold air. A violent upward spiral of air develops, marked by the funnel cloud, and winds near the centre may reach a speed of 200 miles per hour or more. Air pressure drops so rapidly that buildings in the path of the tornado are often destroyed by exploding outwards. Fortunately a tornado is not usually more than 400 yards in width and its average distance of travel is 10 to 15 miles. The chance of any particular place being severely damaged by a tornado, even in an area of frequent tornadoes, is therefore slight.

Tornadoes are most frequent in the United States, especially east of the Rocky Mountains on the plains of the Mississippi basin from Iowa to Texas. There they occur mainly during the spring and summer, moving generally north-east at about 30 to 40 miles per hour. A tornado has been reported only very occasionally in the British Isles.

ANSWER THESE QUESTIONS

1 From what kind of cloud does the funnel cloud in a tornado extend?
2 How wide is the average tornado and how fast does it travel?

25 *Tornado at sea: a waterspout*

When a tornado occurs at sea it is called a *waterspout*. Look at the long, slender funnel cloud of the waterspout in this picture. It extends downwards from a dark cumulonimbus cloud to the surface of the sea, and twists and bends like the funnel cloud of a tornado on land. Another way in which the waterspout resembles the tornado is that there is often a thunderstorm with the cumulonimbus cloud. Near the centre of the waterspout there is a small area of very low air pressure, as there is in a tornado. The violent upward spiral movement of air associated with this low pressure draws up a column of spray from the sea, which can be clearly seen at the bottom of the funnel cloud in the picture. At the same time the movement of air disturbs the surface of the sea around the column of spray and makes it rough.

One kind of waterspout occurs when a tornado which has formed over the land crosses the coast and continues over the sea. Another type forms over the sea and is seen in tropical or sub-tropical regions. The waterspout, like the tornado, has only a short life, lasting up to about half an hour. As the funnel cloud bends and twists, the column of spray finally breaks away from it and the waterspout disappears.

ANSWER THESE QUESTIONS

1 What causes a column of spray at the bottom of the funnel cloud of a waterspout?
2 How long does a waterspout usually last and how does it finally break up and disappear?

26 *Tropical cyclone: a hurricane at 20,000 feet*

A *tropical cyclone* or tropical revolving storm is a system of low air pressure which forms in tropical regions. It is often far more destructive than a tornado because it covers a much larger area and lasts for a longer time. Winds of tremendous strength, sometimes with speeds of 100 to 200 miles per hour, blow round the centre in the same way as they do in a depression: counter-clockwise in the northern hemisphere, clockwise in the southern hemisphere (see end-paper No. 2). In the West Indies and Central America it is known as a *hurricane*. The photograph was taken from a hurricane research aircraft flying at 20,000 feet, and shows the centre, or eye, of one of these great storms. Notice the circular rolls of cloud and the clearness of much of the sky. Here at the eye, which may measure from about 5 up to 20 miles across, the air pressure is very low indeed and the wind is light. Outside the eye is a vast wall of dark cumulonimbus cloud which brings torrential rain, thunder and lightning.

The tropical cyclone moves generally westwards at about 10 to 15 miles per hour, curving round to N.E. in the northern hemisphere and to S.E. in the southern hemisphere. Any place in its direct path has many hours of violent wind and heavy rainfall, followed by the clear calm weather of the eye lasting for perhaps half an hour. Then the storm begins again in all its fury, with the wind blowing from the opposite direction. The tropical cyclone forms mostly on the western sides of great oceans. In the Indian Ocean and the Arabian Sea it is called a *cyclone*; in the China Seas, where it is most frequent, a *typhoon*; off Western Australia a *willy-willy*. It causes immense damage and often loss of life, particularly on the islands and in the coastal districts on the mainland. In recent years such storms have been located and tracked not only by aircraft but also by the use of weather satellites.

ANSWER THESE QUESTIONS

1 How does the weather at the eye of a tropical cyclone differ from the weather elsewhere in the storm?

2 What is the name for the tropical cyclone in the West Indies, the Indian Ocean, the China Seas, Western Australia?

27 *Whirlwind in a dry region: a dust devil in Tanzania*

As you have seen on previous pages, the heating of the earth's surface by the sun causes strong convection currents to develop in the atmosphere. These convection currents may lead to the formation of heavy cumulus or cumulo-nimbus clouds, as condensation of water vapour in the rising air takes place. In a dry region, however, the air contains so little water vapour that heap clouds may not form. Convection currents still develop when rock, or sand, or bare soil is strongly heated by the sun, but they show themselves in a different way. Look at the small whirlwind in the middle distance of the picture. In this part of western Tanzania, only 5° south of the equator, natural vegetation has been largely stripped from the land to make way for crops, leaving a dry, powdery topsoil. The fierce tropical sun has started a convection current, and a whirlwind has developed, raising a column of dust high into the air. This is known as a *dust devil* or *sand pillar*. In the deserts such dust devils are common, and often several may be seen at the same time.

The particles of dust in a dust devil are swept round and round the centre and are occasionally lifted 2,000 or 3,000 feet above the earth's surface. Usually, as with the one in the picture, the column is no more than 100 feet in height. Notice, too, that at the base it is only a few yards across—barely the width of the road. Dust devils generally move along at 5 to 15 miles per hour, though they sometimes reach speeds exceeding 30 miles per hour. Most of them only last for a few minutes.

ANSWER THESE QUESTIONS

1 In a dry region, what often prevents heap clouds from forming as a result of convection currents?

2 What is the usual height of a dust devil? How fast does it generally travel?

28 *Duststorm in the desert: a haboob in the Sudan*

The great wall of dust in the background of the picture is being carried along in a *duststorm*. In the Sudan, where the picture was taken, the storm is called a *haboob*. Such duststorms are well known in the hot deserts of north Africa and south-west Asia, and are a far greater nuisance than dust devils (No. 27). They are not only unpleasant, filling eyes, ears and nose with fine, gritty dust particles: they also seriously reduce visibility, sometimes almost to zero, and make it impossible for aircraft to land or take off. As you see in the picture, the dust rises to immense heights, sometimes to 10,000 feet or more above the ground. A strong wind must blow in order to raise the dust in this way and cause a duststorm. Sometimes the strong wind is associated with cumulonimbus cloud and a thunderstorm, when, for instance, a cold front is crossing the desert. The duststorm moves ahead of the front, and rain falling from the cumulonimbus cloud may evaporate in the hot, dry desert air before it reaches the ground. If the rain does reach the ground, it washes vast quantities of dust particles out of the atmosphere and shortens the duration of the duststorm.

Duststorms, like dust devils, also occur outside deserts in regions where a dry, loose topsoil covers the ground. They have often occurred on the Great Plains of the United States, especially in the 1930s, when the Great Dust Bowl was formed. During those years masses of dust were carried eastwards for hundreds of miles—as far away as the Atlantic coast.

You should remember that sand particles are much larger and therefore heavier than dust particles. This means that sand particles are not raised so easily nor carried so far by the wind as dust particles. A *sandstorm* may develop in the same way as a duststorm and prove equally unpleasant, but it seldom extends above the ground more than 50 to 100 feet.

ANSWER THESE QUESTIONS

1 How may rain affect the duration of a duststorm?

2 How does a sandstorm differ from a duststorm? What is the reason for their difference?

Do You Remember?

See if you remember the meanings of these terms—
then check by referring to the given pages

FURTHER WORK

1 On an outline map of the British Isles and Western Europe draw a typical depression, showing isobars, warm front, cold front, occlusion, and wind arrows (see end-paper No. 2).

2 Describe briefly the instruments you would expect to find in the Stevenson screen at a weather station.

3 Describe the weather changes at a place as the warm front and cold front of a depression pass through it (see end-paper No. 3).

4 Find a copy of the Beaufort scale of wind speeds. Copy the numbers 0 to 12 and beside each write how you would estimate wind speeds in m.p.h.

5 From a reference book or diary find the times of sunrise and sunset on the following dates: spring equinox (about 21st March); summer solstice (about 22nd June); autumnal equinox (about 22nd September); winter solstice (about 22nd December). Work out the hours of daylight on these dates.

6 Describe how radiation fog and advection fog are formed.

7 Write notes on the following: mist, Scotch mist, haze, smog.

8 Draw diagrams of the following types of cloud, showing them at different heights above the earth's surface (low, medium, high): stratocumulus, fair-weather cumulus, altostratus, cirrus.

9 Describe the weather that is usually associated with a cumulonimbus cloud.

10 Both Galileo (1564–1642) and his pupil Torricelli (1608–1647) invented important weather instruments. Find out what you can and write notes about these two inventions.

11 Explain with the help of a diagram how sunlight passes through a raindrop when a rainbow is formed. If necessary consult a book of physics (light).

12 Describe the instrument that is used to measure rainfall.

13 Describe the following and explain the differences in their method of formation: dew, hoar frost, rime, glazed frost.

14 Find out which of the following important European seaports are ice-free all the year and which are not: Antwerp, Rotterdam, Stockholm, Helsinki, Marseilles, Leningrad.

15 Describe the weather in a tropical cyclone. Collect newspaper descriptions of hurricanes and similar storms.

16 Explain with diagrams how sea breezes and land breezes form (see end-paper No. 4).